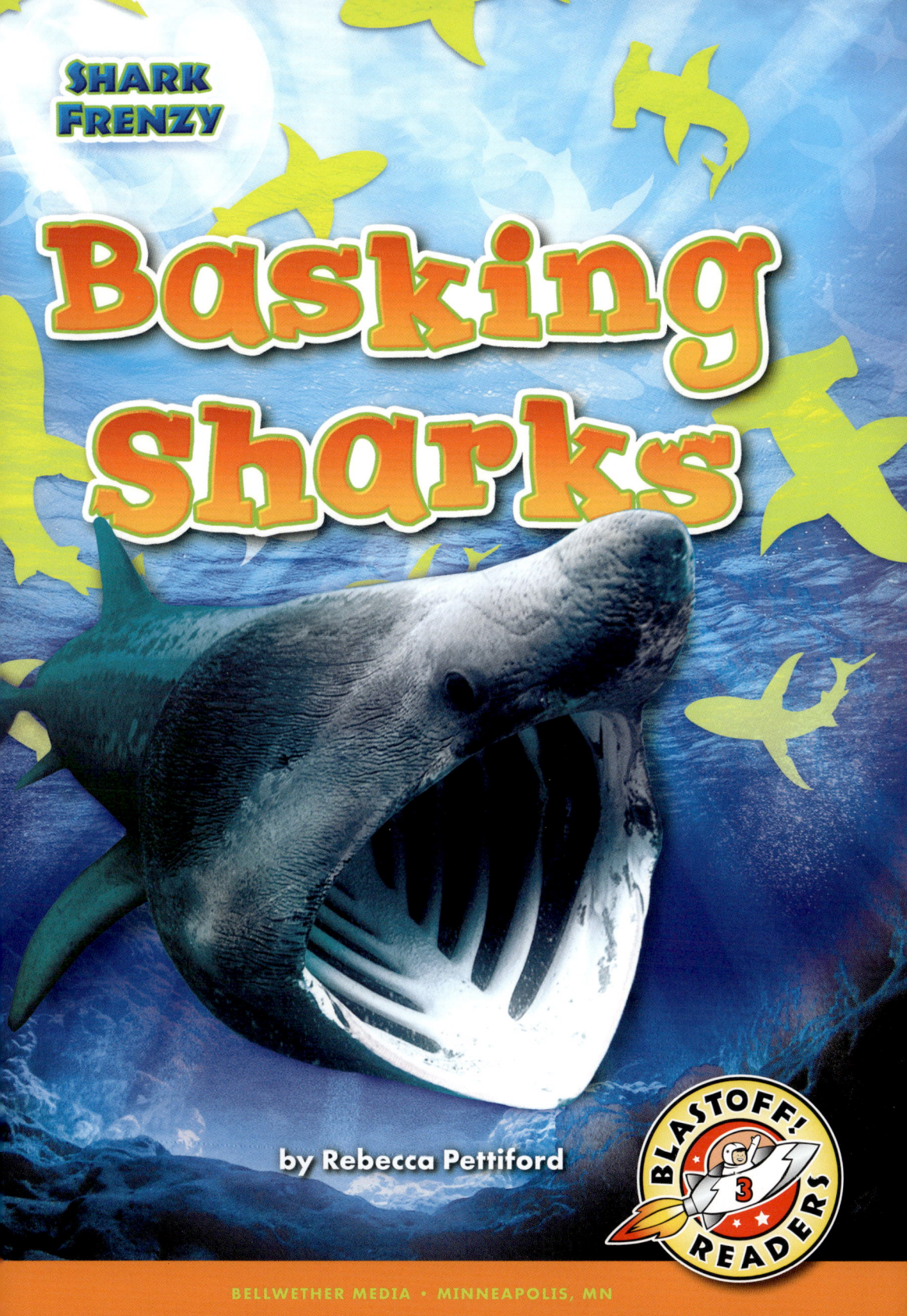

Blastoff! Readers are carefully developed by literacy experts to build reading stamina and move students toward fluency by combining standards-based content with developmentally appropriate text.

 **Level 1** provides the most support through repetition of high-frequency words, light text, predictable sentence patterns, and strong visual support.

 **Level 2** offers early readers a bit more challenge through varied sentences, increased text load, and text-supportive special features.

 **Level 3** advances early-fluent readers toward fluency through increased text load, less reliance on photos, advancing concepts, longer sentences, and more complex special features.

★ **Blastoff! Universe**

   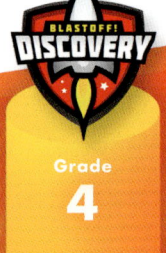

This edition first published in 2021 by Bellwether Media, Inc.

No part of this publication may be reproduced in whole or in part without written permission of the publisher. For information regarding permission, write to Bellwether Media, Inc., Attention: Permissions Department, 6012 Blue Circle Drive, Minnetonka, MN 55343.

Library of Congress Cataloging-in-Publication Data

Names: Pettiford, Rebecca, author.
Title: Basking sharks / by Rebecca Pettiford.
Description: Minneapolis, MN : Bellwether Media, 2021. | Series: Blastoff! readers: Shark frenzy | Includes bibliographical references and index. | Audience: Ages 5-8 | Audience: Grades 2-3 | Summary: "Simple text and full-color photography introduce beginning readers to basking sharks. Developed by literacy experts for students in kindergarten through third grade"– Provided by publisher.
Identifiers: LCCN 2020036786 (print) | LCCN 2020036787 (ebook) | ISBN 9781644874370 (library binding) | ISBN 9781648341144 (ebook)
Subjects: LCSH: Basking shark–Juvenile literature.
Classification: LCC QL638.95.C37 P48 2021 (print) | LCC QL638.95.C37 (ebook) | DDC 597.3–dc23
LC record available at https://lccn.loc.gov/2020036786
LC ebook record available at https://lccn.loc.gov/2020036787

Text copyright © 2021 by Bellwether Media, Inc. BLASTOFF! READERS and associated logos are trademarks and/or registered trademarks of Bellwether Media, Inc.

Editor: Rebecca Sabelko   Designer: Josh Brink

Printed in the United States of America, North Mankato, MN.

Table of Contents

What Are Basking Sharks?	4
Raking Giants	8
Open Wide	14
Deep Dive on the Basking Shark	20
Glossary	22
To Learn More	23
Index	24

What Are Basking Sharks?

Basking sharks are slow-swimming sharks. They live in cool coastal waters all over the world.

They get their name because they feed near the water's surface. They look like they are **basking** in the sun!

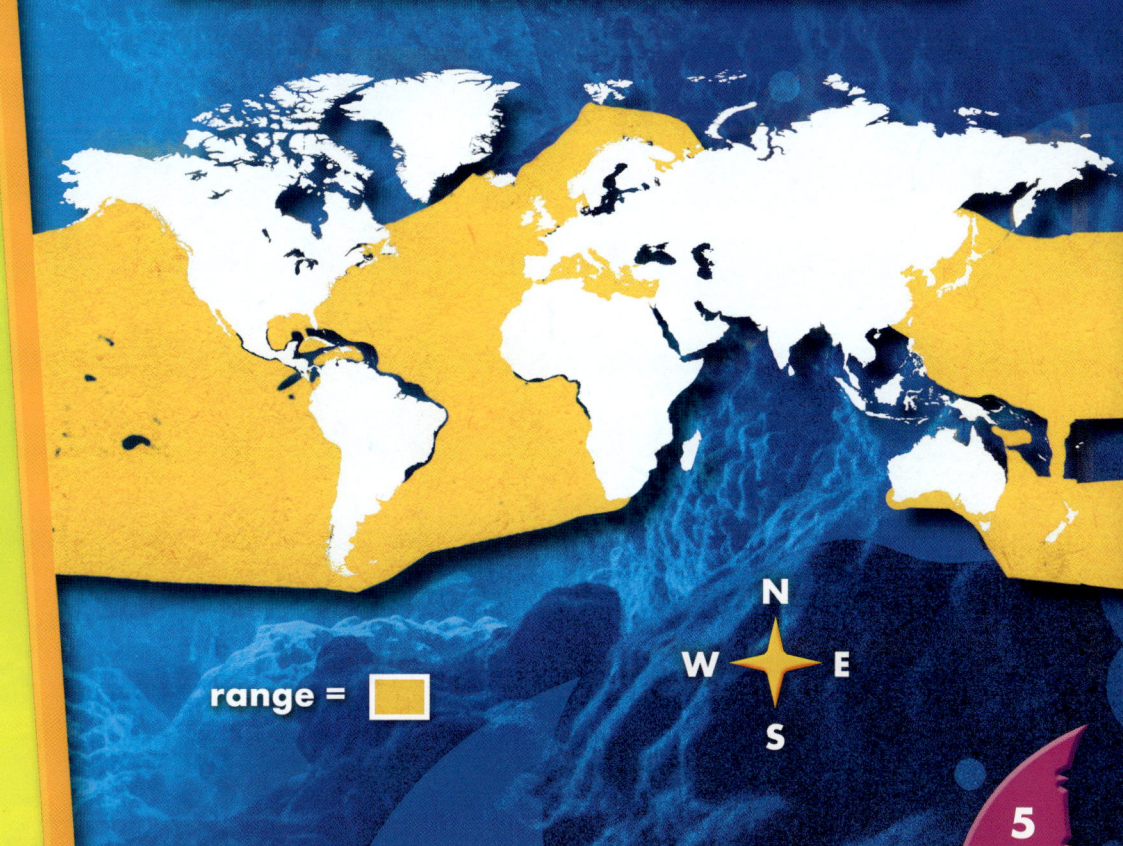

Basking Shark Range

range =

Basking sharks are **endangered**. For hundreds of years, people hunted these sharks for their oil, livers, and fins. Today, overfishing continues.

fishing for basking sharks, 1700s

But people all over the world are working to save these gentle sharks.

Basking Giants

snout

Basking sharks are the world's second-largest sharks.
They grow up to 40 feet (12 meters) long. They can weigh around 10,000 pounds (4,536 kilograms)!

Their long bodies start with their pointed **snouts**. They end with moon-shaped tails.

Shark Sizes

- average human
- basking shark

up to 40 feet (12 meters) long

6 feet (2 meters) long

Basking sharks are covered in toothlike **scales**. They help the sharks move quietly through water.

These sharks have dark backs. Their bellies are dark or white.

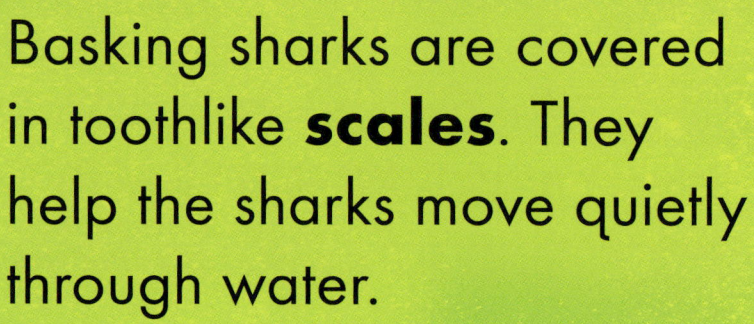

Basking sharks have five large **gills** that nearly circle their heads. Their gills have **gill rakers**. Gill rakers **filter** water and trap food.

gills

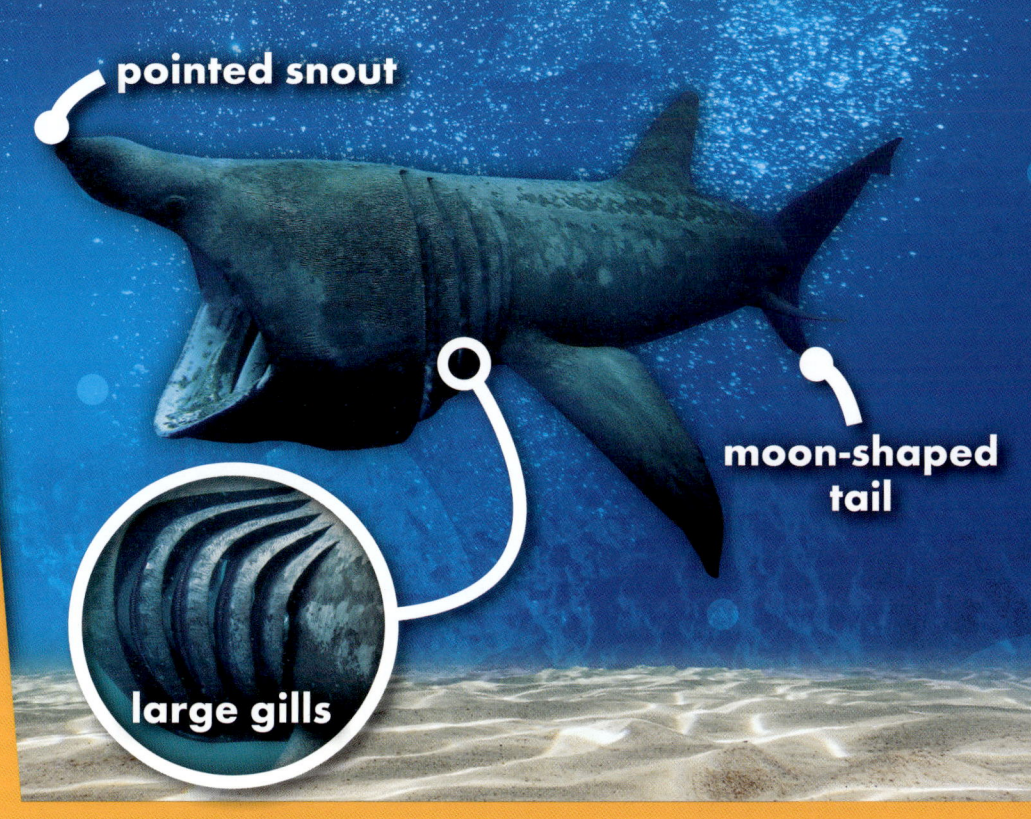

Identify a Basking Shark

- pointed snout
- moon-shaped tail
- large gills

They shed their gill rakers in winter. New gill rakers grow in the spring.

Open Wide

Basking sharks are **filter feeders**. They swim slowly with their wide mouths open.

Plankton, fish eggs, and shrimp flow into their mouths. The gill rakers trap the food.

Basking Shark Diet

plankton

fish eggs

shrimp

Basking sharks **migrate** with the seasons. In summer, they travel to northern waters to feed.

In winter, they swim to warmer areas to **mate** and have their young.

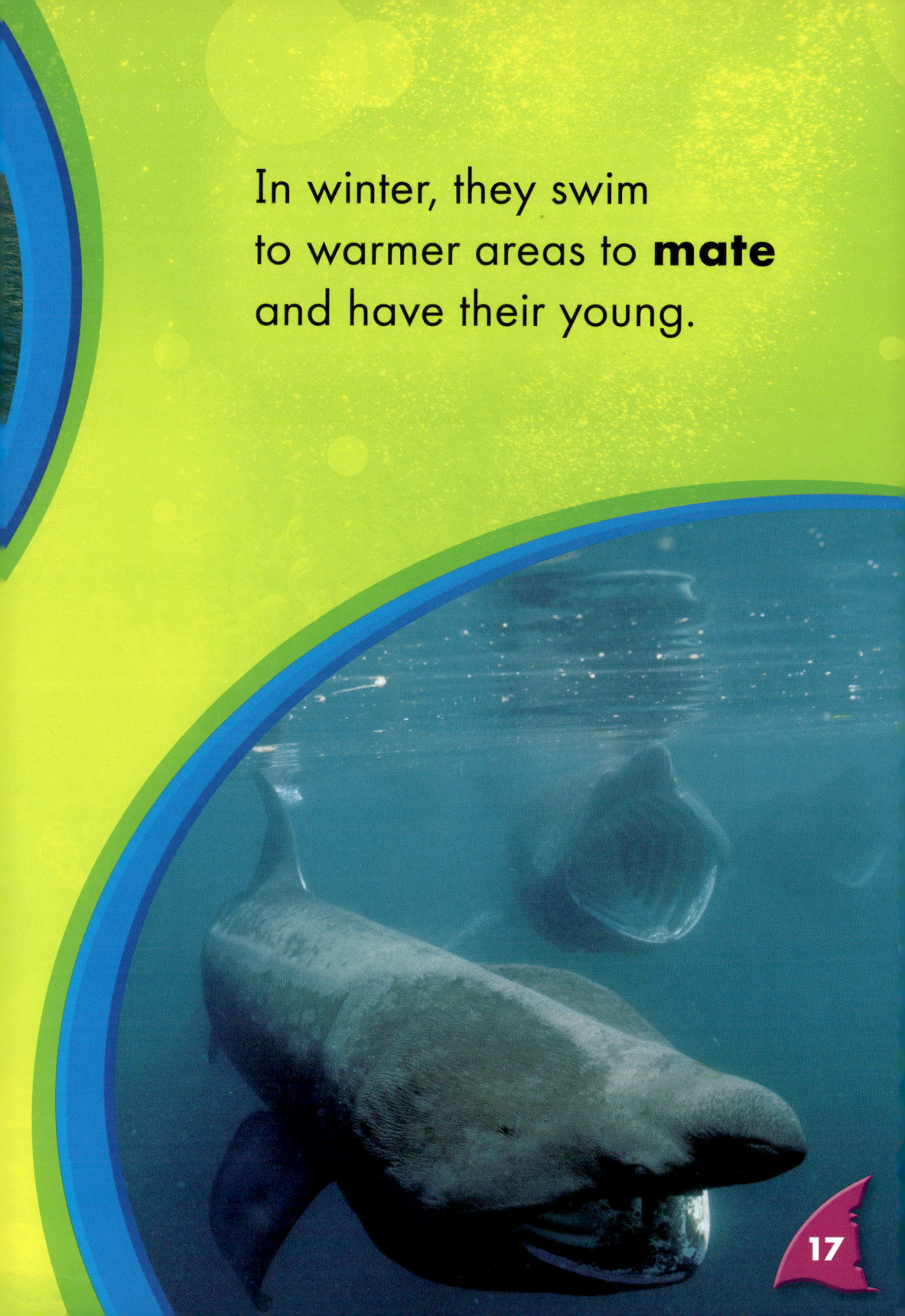

Basking sharks do not have many **predators**. Killer whales and great white sharks sometimes hunt these slow swimmers.

These gentle giants have little to fear in the ocean!

Deep Dive on the Basking Shark

moon-shaped tail

LIFE SPAN:
up to **50 years**

LENGTH:
up to **40 feet (12 meters) long**

WEIGHT:
up to **10,000 pounds (4,536 kilograms)**

DEPTH RANGE:
0 to 6,562 feet (0 to 2,000 meters)

Glossary

basking—relaxing in a bright and warm place

endangered—at risk of dying out

filter—to pass a liquid through a material to trap small parts in the liquid

filter feeders—ocean animals that take in many small pieces of food at one time

gill rakers—comb-like structures in gills that trap food

gills—parts that help sharks breathe underwater

mate—to join together to make young

migrate—to travel from one place to another, often with the seasons

plankton—ocean plants or animals that drift in water; most plankton are tiny.

predators—animals that hunt other animals for food

scales—small plates of skin that cover and protect an animal's body

snouts—the noses of some animals

To Learn More

AT THE LIBRARY

Murray, Julie. *Basking Sharks*. Minneapolis, Minn.: Abdo Zoom, 2020.

Pettiford, Rebecca. *Whale Sharks*. Minneapolis, Minn.: Bellwether Media, 2021.

Waxman, Laura Hamilton. *Basking Sharks*. Mankato, Minn.: Amicus High Interest, Amicus Ink, 2017.

ON THE WEB

FACTSURFER

Factsurfer.com gives you a safe, fun way to find more information.

1. Go to www.factsurfer.com.
2. Enter "basking sharks" into the search box and click .
3. Select your book cover to see a list of related content.

Index

backs, 10
bellies, 10
bodies, 9
color, 10
deep dive, 20-21
filter feeders, 14
food, 12, 15
gill rakers, 12, 13, 15
gills, 12, 13
hunt, 6, 18
mate, 17
migrate, 16
mouths, 14, 15
name, 5
ocean, 19
overfishing, 6
predators, 18
range, 4, 5

scales, 10
seasons, 13, 16, 17
size, 8, 9, 19
snouts, 8, 9, 13
status, 6
swim, 4, 14, 17, 18
tails, 9, 13
waters, 4, 5, 10, 12, 16
young, 17

The images in this book are reproduced through the courtesy of: Doug Perrine/ Minden, front cover (hero); Mark William Kirkland, p. 3; Nature Picture Library/ Alamy, pp. 4, 8, 10, 13 (call out), 14, 19; Chronicle/ Alamy, p. 6; Alex Mustard / 2020VISION / Alamy, p. 7; Charles Hood/ Alamy, pp. 10-11, 12; Paulo Oliveira/ Alamy, pp. 13, 18, 20-21; FLPA/D P Wilson/ agefotostock, p. 15 (top left); valda butterworth, p. 15 (top right); Mati Nitibhon, p. 15 (bottom); George Karbus Photography/ agefotostock, p. 16; Charles Hood/ Newscom, p. 17; Martin Prochazkacz, p. 23.